Already There

By

Robert Charles Lewis

This book I am dedicating to my son Daniel Robert Lewis.

He is the ONLY person who encouraged me, and it was simply by saying, "Dad, you can do it."

Books By: Robert Charles Lewis

1. Beyond The Infinite
2. Poems of Multiple Meanings
3. Poems of Thoughtful Faith
4. Writings of Thoughtful Joy
5. Words of Truth
6. Developmental Peace
7. Truisms of Life
8. Continuing Life
9. Waterways
10. Hope
11. Tests of FAITH
12. The Quest For Finite Pi
13. Writings of Muse
14. The Socioeconomic Impact of Injustice
15. Together Again
16. Quips, Haiku, and Other Poems
17. A Book of Songs
18. The Simplicity of the Bible
19. Incremental Math

All books are available thru Amazon.com. Enter book title AND full name of author (Robert Charles Lewis).

Foreword

In this book I am taking a fresh look into mathematical principles.

In so doing I will provide easy to comprehend solutions for otherwise very complicated equations.

Contents

Chapter 1

<u>The Essence of Time</u>

Time can be defined, measured, and easily understood.

The purpose of what I am about to write is to that end.

With a thorough understanding of time all currently unsolved questions, problems, equations, and even philosophies about the very existence of a God who is a Creator, is omnipresent, and who has NO beginning and NO end can be accounted for.

Now, to start with, a key concept of time that is currently accepted as a standard, absolute truth, is our use of minutes, hours, days, seconds, years, months, as a mathematical rule-of-thumb.

For example, the use of 'miles-per-hour' causes mostly false understandings in solving, and even being able to solve all such expressions of time and also rates of velocity.

Time is simply a relationship between a point and any and all number of points ranging from zero to the infinite.

For example, time on earth is expressed as a relationship between two points, namely the earth and the sun. Noon and midnight, being 12 hours apart, are an example of that relationship.

If one were to add another point to that example, such as using the moon, then the subsequent calculation of time would result in an expression of time far different than what mankind currently uses.

I clearly expressed in previous ways that time as we know it is merely a "physical, positional relationship between the earth and the sun." However, time can be a physical, positional relationship between any number of points from zero to infinity.

By now I hope that you, the reader, are starting to understand the shortcomings of our current expression of time. The implications of the many, many benefits to mankind that will result in the use and applications of updated knowledge continues to drive me personally as the author to strive to explain myself clearly.

Now, to mention the "points" used in the calculation of time, which I will now refer to as "relative-time:"

Each point can be totally unique to any and all points.

Also, each point can be exactly the same in all ways.

Any and all combinations of motion and/or non-motion can be used.

Time can be expressed as ALL shapes, from, as an example, a triangle, square, pentagon, hexagon, etc. ALSO, time can be a sine-curve on one side, a secant on the other, and a tangent on another, etc.

Time therefore is the key to understanding space itself.

In ending, as a proper understanding of all the elements of any situation, whether a thought or a written expression, are needed to arrive at a correct resultant answer, I believe that time as I understand it will be the key for me to personally solve all 7 of the Millennium Prize problems. And with the GOOD LORD Jesus Christ's blessing, I will.

Chapter 2

<u>Navier-Stokes existence and smoothness</u>

My solution to this problem I derived by first addressing the two conditions that are excluded from the equations, namely zero velocity and infinite velocity.

It dawned on me that, very simply put, an object at zero velocity is, as far as being three dimensions and time, "already there" in its trajectory.

Time itself is both a "relative-movement," as well as a "relative-position."

I also took into account the definition of velocity as it includes a measurement of time.

Therefore, another explanation of my "already there" is again very simple to understand and prove mathematically because a stationary object, meaning one with absolutely no velocity, has a time consideration of zero.

Now, an object at a truly infinite velocity is also "already there" at each and every point of its trajectory.

It also has a time consideration of zero, just as a stationary object has. This is true because at a truly infinite rate of velocity the object is "ever-present," meaning present simultaneously at each and every point of its trajectory, and thusly it exists in timelessness, just as does a stationary object.

Now I will discuss objects in motion. Regardless of what the velocity of an object is, it also is already-there at each and every point along its trajectory.

A summary of my answer and verbal proof of the Navier-Stokes existence and smoothness problem is that a stationary object has a scalar pressure of ZERO, as also do objects of ANY and ALL rates of velocity, because they are now by my written explanation shown to be present at each and every point (already there).

All scalar pressures are a CONSTANT and also there is absolutely no turbulence in any case.

Chapter 3

A Proof Of 'Already There'

1. Zero Velocity (Stationary)

A. Always at <u>same</u> point.

B. 0-V-0-T 1 point.

C. V x T = P

D. 0 x 0 = 1P

2. Infinite Velocity

A. Always at <u>ALL</u> points.

B. ∞-V-0-T, ∞ points.

C. V x T = P.

D. ∞ x 0 = 1P

3. Steady Rate Velocity

A. Always at <u>EACH</u> point (along a trajectory).

B. ∞-V-0-T, ∞ points.

C. V x T = P.

D. ∞ x 0 X1 = 1P

0 x 0 x 1 = 1; ∞ x 0 x ∞ = 1; ∞ x 0 x 1 = 1.

<u>ALL</u> rates are the same as to velocity, time, and point.

A. Let "IV" stand for "infinite velocity."

B. Let "OV" stand for "zero velocity."

C. Let "RV" stand for "rate of velocity."

D. Let "T" stand for "time."

E. Let "P" stand for "point."

F. Let "L" stand for "location."

A. OV x T x P = L.

B. IV x T x P = L.

C. RV x T x P = L.

D. OV = IV = RV.

End of proof.

Chapter 4

P versus NP

Very simply put, **P=NP**.

Regardless of the complexity of any and all problems in computer science, they are readily both solved and verified by a properly programmed computer. The only current trouble that computers have is in the fact that they use out-dated forms of processing data which, in turn, lead to the erroneous belief of many that computers are incapable of providing solutions to all problems.

To make myself clear on this point, the day is coming when computers will be able to provide simultaneously-instantaneous solutions and verifications of problems.

I have several reasons that I will now write to explain my claim in the previous paragraph: Number 1. We are all children of the same Higher Power. Our maker is all-knowing, and knows the end from the beginning. This means that all answers are known to our maker from the instant they are formed. We, as children of our maker, each and every one has been made in a unique way and have gifts, such as abilities, that are from the Highest. We need to communicate both with our Maker, and also with our other created "siblings." GOD is LOVE and wants us to live in peace and harmony with all. In so doing, our Heavenly Maker will increase our knowledge and wisdom, which in turn when applied correctly, leads to blessings for all mankind.

Number 2. In my book titled "Incremental Math," I discussed the problems of today's mathematics, in that we use "base 10" to write and comprehend solutions of "numeric values." Also, computers use "base 2," and with these types of conditions, these lead to many currently unsolvable problems encountered in all professions that use numerical expressions to, say, reason with. In short, there exist "math-languages" that can solve the entire spectrum of problems, such as those encountered in computer science.

As for the time spent on problems in order to find a solution, I discussed time itself in three (3) of my books. In my book titled "Beyond the Infinite," I wrote, "The past and the future are ever-present." In my book titled "The Quest for Finite Pi," I discussed Time in all of chapter 1.

To summarize this 2nd reason that computers will one day give simultaneously-instantaneous solutions and verifications to all problems presented to it is that we children of the Most High God are all "expressions" of Him, our Heavenly Father, and He is ever-present, immortal, and all-knowing. Therefore we also are always with Him in all ways that He exists.

Even though our "flesh and blood" mortal bodies grow old, and return to the earth, it is our Spirit that is eternal. GOD IS A SPIRIT and ALL Spirits return to Him, and again, GOD loves all His children.

With GOD ALL things are possible, and here-in is an understanding of why there are so many short-falls in mathematics, physics, science, philosophy, etc. It is because mankind has fallen away from both knowing GOD's WORD to we His children, and then also because mankind has not remained steadfast in relying on the key truths of Holy Scripture, that WILL lead to 100% understanding of all that exists.

Number 3. For now, my ending is that all of us make mistakes, and we must not let those stop us from making true advances in our callings and professions. Teamwork is a key ingredient to success, and to learn from others can be both individually and mutually beneficial.

Chapter 5

Thoughts I Built On

Navier-Stokes existence and smoothness

A very simple viewpoint that is overlooked in theoretically understanding these set of problems is that both a liquid and a gas are nothing more than a heated SOLID. Ergo, this means STP prevails as the key to completely understanding all the aspects of fluid mechanics, even turbulence.

Atomic weight is also to be taken into consideration in order to understand how different gases, liquids, and solids (also semi-solids) act under diverse situations. [I was interrupted at about 4:15 AM on 1-14-17 to make a cup of coffee for John Atkins, and it is now Sunday, January 22nd, 2017 as I write these details. Also, I decided to finish my 19th book, titled "Incremental Math," which I accomplished. It is now 6:12 AM.]

Navier-Stokes existence and smoothness

A very simple viewpoint that is overlooked in understanding these problems is the fact that both a liquid and a gas are indeed forms of a heated solid. Ergo, this means that 'STP' is one of the keys to a correct solution to all aspects of fluid mechanics, including turbulence.

Another consideration is the atomic weight of each type of material that a problem is addressing, because diverse elements will, of course, act in diverse manners.

Another thought to take into account is the very definition of velocity itself because it involves time.

Time is both a 'relative-movement' and also a 'relative-position'. For example, at ZERO velocity an object is stationary, in other words, 'already-there'. Likewise, at an INFINITE velocity an object is 'already-there' at an infinite distance from its stationary point, its origin. Now it follows that it also holds true that the same object is 'already-there' at ALL POINTS of its trajectory. It also holds true that from any point of an object's trajectory (such as a mid-point, for example), the object is 'already-there' in opposite directions at all points.

With this analogy, it can be mathematically derived that in three space dimensions and time, with any initial velocity field, there exists a vector velocity and a scalar pressure field, which are both smooth and globally defined, that solve all the Navier-Stokes equations.

And as a reminder, a velocity of ZERO is equal in time to a velocity of an INFINITE rate, because the INFINITE is actually a state of 'timelessness,' ergo ZERO time equals INFINITE time.

Unknowns are:

1. Velocity – "V" (x, t).

2. Pressure – "P" (x, t).

Note: In 3 dimensions, there are 3 equations and 4 unknowns (3 velocities + 1 pressure) ((Another equation is needed)).

In space, fluids are not within their element, ergo, there are more variables to reckon with.

1. Friction

   A. slows.

   B. causes turbulence

2. Magnetism

   A. repel/attract

   (1) slows

   (2) turbulence

3. When fluid is the same material

as pipe, . . . etc, . . . etc.

4. Temperature

   A. Fluctuations

   (1) environmental

   (2) frictional

5. Adhesive/Cohesive Properties

As time is 'relative-movement,' then <u>ZERO</u> time means that an object is stationary, and 'already-there.'

Likewise, at a TRULY INFINITE rate of velocity, then an object is stationary and 'already-there' along all points of its trajectory. Therefore . . . (later) . . . at all points (increments), along the trajectory of an object in motion at an infinite rate of speed, it follows that the object is present in its original form, and is ever-present along its trajectory (God is omnipresent also).

Now, using this TRUTH, then, as I stated in #2, P=NP because a computer's speed of operation (a type of velocity) can be, thru proper

"thinkology," increased to the speed of light using, as an example, fiber optics to assemble it with, instead of copper wire.

Eventually, all answers thru the use of computers can be expressed simultaneously, instantaneously, using different branches of mathematics to equate situations. (Also laser communications are rapid.)

Already There

Almost There

Already There at Each Point of Trajectory

Ergo, pressure is a constant at all velocities.

Time as a;

A. Linear existence, and/or;

B. Linear non-existence.

(Regardless of curvilinear, or not)

Note: Ergo, another way to describe a curve, straight line, etc, is the angle(s) of incidence of a set of points.

All shapes are simply a CURVED STRAIGHT LINE!!!

P versus NP

Very simply put, P=NP.

A foundational PROOF is that GOD knows all the answers, even before they are formed in mankind's worldly mathematical equations, and HE WILL show to ALL those who trust in Him, and are so inclined to base their lives in search of specific areas of thought, and ask in Christ Jesus' name, the answers to all things.

Regardless of complex equations created by mortal minds, there are very, very, very easy to understand formulas and equations that contain answers to humanity's every mathematic (etc) problem.

In ending, as a furtherance of understanding the truism that P=NP, be it understood that the day will come when answers will be evident instantly, regardless of content or complexity.

Colour is bound to preserve the beauty of our Maker's creation, and is a proof in itself that a Higher Power exists.

Byword

PSALM 16:8. I ALWAYS PUT GOD FIRST AND I WILL NOT BE MADE TO THINK OTHERWISE.

LUKE 23:31. WHEN NO MIRACLES ARE BEING DONE THEN LIFE IS MUCH HARDER FOR CHRIST'S FOLLOWERS.

REVELATION 21:1. I HAVE BOTH SEEN AND DWELLED ON THE NEW EARTH.

DEUTERONOMY 8:1. OBEY ALL OF GOD'S COMMANDMENTS!!!

www.ingramcontent.com/pod-product-compliance
Lightning Source LLC
Chambersburg PA
CBHW081652220526
45468CB00009B/2624